Who Pooped in the Park?

in the Park

Written by Gary D. Robson
Illustrated by Robert Rath

Death Valley National Park

FARCOUNTRY PRESS

HELENA, MONTANA

To my in-laws, Walter and Barbara.
I finally did a Death Valley book!
- Gary

For Lucy and Thomas, my poop experts.
- Robert

ISBN 10: 1-56037-403-9
ISBN 13: 978-1-56037-403-9

For more information on our books,
write Farcountry Press, P.O. Box 5630, Helena, MT 59604;
call (800) 821-3874; or visit www.farcountrypress.com.

Book design by Robert Rath.
Created, produced, and designed in the United States.
Printed in China.

12 11 10 09 08 07   1 2 3 4 5 6

Library of Congress Cataloging-in-Publication Data

Robson, Gary D.
Who pooped in the park?. Death Valley National Park / written by Gary D. Robson ; Illustrated by Robert Rath.
p. cm.
ISBN-13: 978-1-56037-403-9
ISBN-10: 1-56037-403-9
1. Animal tracks--Death Valley National Park (Calif. and Nev.)--Juvenile literature. I. Rath, Robert ill. II. Title.
QL768.R617 2007
591.9794'87--dc22
2007028167

"Are we there yet?" Michael squirmed in the back seat. "We've been driving through the desert forever!"

"Hang on," said Dad. "We're almost there."

3

Mom added, "Once we meet up with Sara's family, then you can get out and walk."

Sara was Michael's best friend back home, and they were very excited when they learned their families were both going on vacation in Death Valley National Park.

"Oh, he doesn't really want to get out of the car," teased Emily, Michael's big sister. "He's scared he'll be eaten by a mountain lion!"

"That's enough, Emily," said Mom. "Nobody's getting eaten by anything."

Michael was excited about their trip to Death Valley National Park, but Emily was right. He was a little scared.

"In my wildlife book, the mountain lions do look kind of scary," he admitted.

"Don't worry, Michael," said Dad. "Mountain lions are scared of people, too. They are very rare in Death Valley National Park. We probably won't even see one."

"But that doesn't mean we won't learn about them," Mom said with a smile.

the STRAIGHT POOP

Never hike by yourself. Mountain lions almost never bother people hiking in groups.

Finally, after the long drive, they arrived at the campground. Emily spotted Sara right away.

"There's Sara," Emily said. "Now Michael can bug *her* instead of me."

Mom invited Sara to hike with them, but asked her to leave her dog, Rex, with her parents. "Dogs aren't allowed on the hiking trails, Sara," she said, "because they might scare the wildlife."

"Okay," said Sara. "He wouldn't like this hot weather anyway."

"Then why bring him here?" Michael asked.

"Not all of Death Valley is this hot. It's much cooler in the mountains," Sara replied. "Rex loved the mountains where we were yesterday."

the STRAIGHT POOP

Always wear sunscreen and carry drinking water when hiking in Death Valley!

Scotty's Castle ●

DEATH
VALLEY
NATIONAL
PARK

Furnace Creek ●
(hottest temperature ever
recorded in the U.S.)

Panamint Springs ●

NEVADA

Badwater ●
(the lowest elevation
in North America)

CALIFORNIA

Ashford Mill ●

"Death Valley is huge," said Mom. "Almost as big as the whole state of Connecticut. It has mountains, deserts, and creeks. And it may not look like it, but it has a lot of animals, too!"

the STRAIGHT POOP

The lowest point in North America is in Death Valley National Park. It also has one of the highest points.

11,049 feet
above sea level
● Telescope Peak

Sea Level

Badwater ●

282 feet below sea level

"Animals?" said Michael, looking around at the barren desert.

"We may not see a lot of animals down here in the desert, but we'll see their sign," Dad answered.

"Sign? Like a sign at the zoo?" Michael asked.

DESERT BIGHORN SHEEP

COYOTE

ROADRUNNER

"When Dad says 'sign,' he means clues that an animal has left behind," explained Mom. "Like tracks in the sand."

"Or piles of poop!" added Sara. When everybody looked at her, she added, "The ranger explained it to us."

Emily spotted strange marks she thought might be an animal sign. But she wasn't sure what it was.

"This is one of the coolest snake tracks you'll ever see," Dad told her. "It's from a sidewinder rattlesnake."

13

"A rattlesnake?" Michael yelled. "Where?"

"It's okay, Michael, this is just *sign* that a rattlesnake has been here. It's not still nearby," Mom told him.

the STRAIGHT POOP

If you see a rattlesnake, just stay away. If you don't bother it, it won't bother you.

Emily looked at the odd twists in the sand. "How could a snake make tracks like these?"

"Different snakes move their bodies in different ways to travel around," Dad told her. "Sidewinders twist their bodies over the loose sand to get from one place to another."

"*Ewww*, what's this?" asked Sara, looking at a stringy piece of poop on the ground.

"This," Mom answered, "is rattlesnake scat."

"*Scat*? What's scat?" asked Michael.

"Scat is the poop that Sara was talking about," said Emily. "That's what scientists call it."

Michael put his hands on his waist, looked around, and asked, "Is there anything out here in the desert besides snakes and snake poop?"

"Sure," answered Dad. "Look!"

There, running across the sand, was a small, brown bird.

"A roadrunner!" Mom said.

"Do coyotes really chase them, like in the cartoons?" Michael asked.

"Coyotes eat roadrunners, and roadrunners eat snakes—if they can catch them," answered Mom.

the STRAIGHT POOP

Roadrunners don't really make a meep-meep sound, as in the cartoons. They coo like a dove.

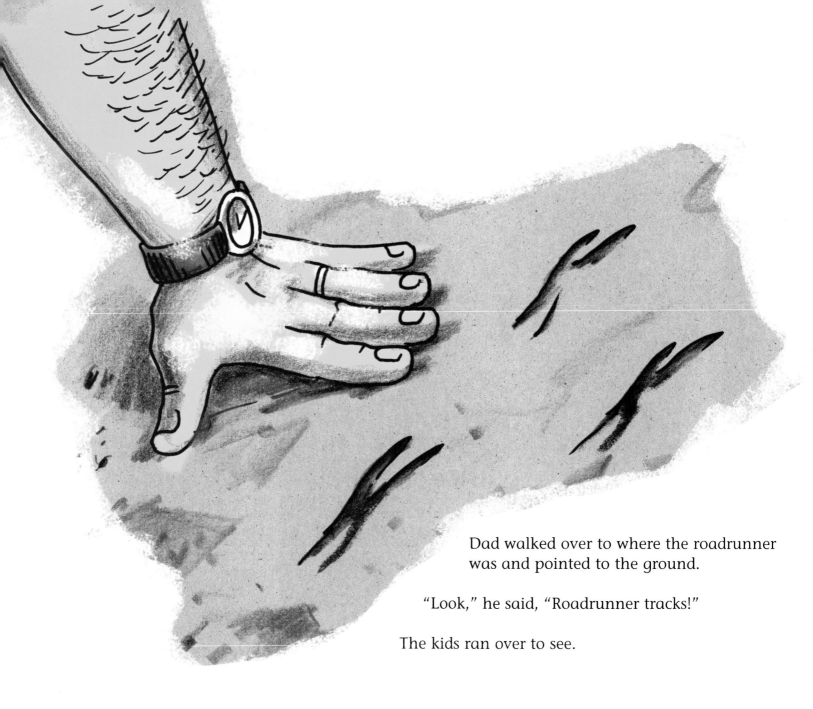

Dad walked over to where the roadrunner
was and pointed to the ground.

"Look," he said, "Roadrunner tracks!"

The kids ran over to see.

"Was there a coyote around here, too?" Michael wanted to know. "Did it leave any tracks?"

"I think I found some coyote tracks over here," yelled Mom. "Come see."

"And that's not all," said Dad. "There's coyote scat, too."

"Wow," said Sara. "I saw bobcat tracks yesterday, but no scat."

the STRAIGHT POOP

You can tell coyote scat from dog poop because coyote scat usually has bits of hair and bone in it.

COYOTE

BOBCAT

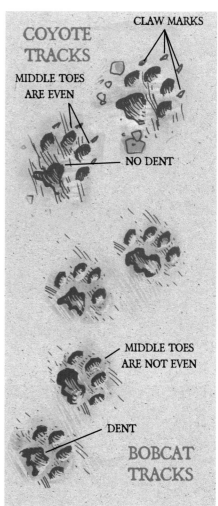

COYOTE
TRACKS

CLAW MARKS

MIDDLE TOES
ARE EVEN

NO DENT

MIDDLE TOES
ARE NOT EVEN

DENT

BOBCAT
TRACKS

"See how these tracks show claw marks, Sara?" Dad said.
"Did you see the claw marks in your bobcat tracks?"

"No," said Sara after she thought for a moment.
"And they were bigger than this, too."

"Bobcats can retract their claws, like a housecat," Dad
explained. "The claws don't show in tracks very often."

"Look at these tracks here," said Emily. "The dirt's pushed up behind them, and the tracks in front of these are spread apart a long way."

"The coyote was crouched here, and launched itself forward hard to chase something," said Mom.

the STRAIGHT POOP

Animal tracks can give you clues about what animals were doing when they left the tracks.

FRONT TRACKS

COTTONTAIL RABBIT TRACKS

BACK TRACKS

"More tracks!" said Michael. He was having fun finding clues. "These are a lot smaller."

"These tracks are from a rabbit," said Dad, "and a coyote was chasing it."

"These aren't as long as the rabbit tracks we saw in the mountains," Sara added. "Was this one a baby?"

"You probably saw jackrabbit tracks up in the mountains, Sara," explained Dad. "Down here in the desert, these are probably cottontail rabbit tracks."

BLACK-TAILED JACKRABBIT

DESERT COTTONTAIL RABBIT

the STRAIGHT POOP

Jackrabbits aren't really rabbits! They're a closely related animal called a hare. Jackrabbits are bigger and skinnier than rabbits, and they have longer legs and longer ears.

"I know who pooped *here*!" said Michael.

"Who?" Sara asked him.

"The rabbit!" said Michael. "It looks just like the poop I have to clean out of Fluffy's cage at home."

the STRAIGHT POOP

Rabbits eat their own scat! They do this to get as much nutrition from the food as they can. The little brown balls are scat that's already been through the rabbit twice.

"I'm getting hot," said Michael, sitting down on a rock.

"And hungry, too," added Emily, plopping down next to her brother.

"Let's go back to camp and get some lunch, and then we can go up to a cooler place in the hills," suggested Mom.

After lunch, they drove to a
hiking trail in the mountains.

"It looks really different
up here," Michael said.

"Yeah. Even the tracks look different," said Emily.

"I told you so," replied Sara.

"Those tracks are *weird*," Michael added.

the STRAIGHT
POOP

You won't see badger scat very often because they usually poop underground in their burrows.

"You found badger tracks," said Dad.

"When a badger walks, the front and back footprints overlap," added Mom. "That's why they look so strange."

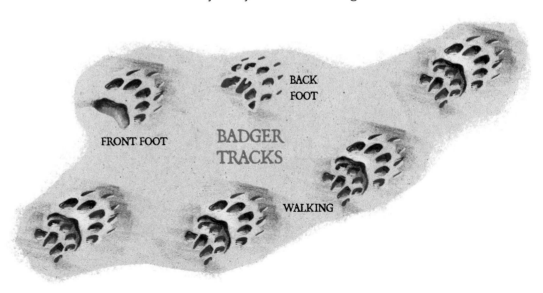

BACK
FOOT

FRONT FOOT

BADGER
TRACKS

WALKING

As they walked along the trail, they came to a creek.

"A clue!" said Sara, who spotted a shape in the mud.

"Those tracks are from a bird called a grebe," Mom told her. "They live near water, like ducks, but their feet are very different."

Even deserts can have streams, rivers, and lakes. It's the lack of rain that makes it a desert, not the lack of water.

GREBE TRACKS

DUCK TRACKS

WEBBING AROUND EACH TOE

WEBBING CONNECTS TOES

DUCK AND GREBE SCAT

the STRAIGHT POOP

Grebe scat and duck scat look alike. You have to look at the tracks nearby to figure out who pooped!

33

A pied-billed grebe burst out of the deepest part of the pool, making Michael jump.

"Wow! Where did *that* come from?" asked Michael.

"We probably scared it when we walked up," said Dad. "It was hiding."

the STRAIGHT POOP

Unlike ducks, grebes are more likely to dive underwater when scared than to fly away.

Farther down the trail, Dad asked, "Does anyone know what animal made these tracks?"

"I saw tracks like these at home!" Emily said. "They're from a deer, right?"

"Close," Dad told her. "These tracks are from a bighorn sheep."

"Bighorn sheep live high up in these mountains—sometimes in big herds," he told them.

the STRAIGHT POOP

Desert bighorn sheep have hooves that are hard around the edges and soft in the center. This gives them better traction, to help them climb on rocks.

BIGHORN SHEEP TRACKS

Sara saw a pile of scat that looked like the rabbit poop they saw earlier. "Was there a bunny hanging out with these bighorn sheep?" she asked.

"No," Mom said with a smile. "Even though sheep and rabbits look really different, their scat looks very similar. This scat is from a bighorn sheep."

BIGHORN SHEEP SCAT

BIGGER PIECES

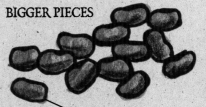

SHAPED LIKE
A JELLYBEAN

JELLYBEANS

RABBIT SCAT

SMALLER PIECES

ROUND LIKE
A BALL

Michael called out from down the trail, "I think I found another coyote track."

"That's not a coyote track," said Sara, catching up. "It's from a mountain lion. Remember what Dad told us about the claw marks showing?"

"What?" Michael gasped. "Is the mountain lion around here somewhere? Maybe we should go back to the car…"

"It's okay," said Mom. "That mountain lion is long gone by now."

"We're very lucky," said Mom. "There aren't many mountain lions around here, but we got to see mountain lion sign."

"Yes," Dad added. "Most animals leave sign accidentally, but this was left on purpose."

"On purpose?" Michael looked puzzled.

"This is called a 'scrape.' The mountain lion scraped all this stuff in a pile and peed on it so other mountain lions would know he was here."
"Ewww!" said Emily.

the STRAIGHT POOP

Mountain lions have different names in different parts of the country. They're also called panthers, painters, cougars, pumas, and catamounts.

41

"Is this mountain lion scat?" asked Michael, still a little scared.

"It sure is," answered Dad.

Michael looked closer. "It has hairs and bits of bone in it, just like the coyote scat," Michael pointed out. "That means they eat other animals."

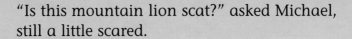

the STRAIGHT
POOP

Mountain lions may be the biggest cat in America, but they still bury their scat just like a housecat.

Emily laid her hand next to the track.
"This mountain lion must be big," she said.

"That's right," Mom said.
"A mountain lion weighs as
much as I do, and a big one
can weigh more than Dad!"

As they ate dinner that night, everyone talked about how much fun they had.

"We didn't see very many animals," said Emily, "but it seemed like we did."

Everyone laughed when Michael said, "And I didn't get scared once!"

45

TRACKS and

SIDEWINDER RATTLESNAKE

Their tracks look like several curvy lines, with space in between; the space is where the snake lifts its body off the ground as it "winds" across the sand.

Scat is long skinny cords.

COYOTE

 FRONT

BACK

Coyote tracks look like dog tracks. They have four toes on each foot. Their claw marks usually show above each toe.

Scat is very dark in color and can contain bits of seeds or hair. Scat is tapered on the ends.

DESERT COTTONTAIL RABBIT

FRONT

BACK

They have four toes on each foot. Because their feet are furry, their toes don't really show up in the tracks. Their back feet are much longer than their front feet. Their claw marks sometimes show in their tracks.

Scat is a pile of small brown balls.

BLACK-TAILED JACKRABBIT

FRONT

BACK

Their tracks are like the tracks of the desert cottontail rabbit, but the front tracks are smaller, and back tracks are longer and thinner.

Scat is a pile of small brown balls.

BADGER

FRONT

BACK

They have five toes on each foot. Tracks show marks from their very long front claws. The claw marks are connected to the toe marks.

Scat is similar to poop from a small dog, with blunt ends, not tapered like a coyote.

SCAT NOTES

PIED-BILLED GREBE

Tracks show three large toes with fringe of webbing around each toe.

Scat is in cords about as big around as a pencil.

DUCK

Tracks show webbing between each toe.

Scat is in cords about as big around as a pencil.

DESERT BIGHORN SHEEP

FRONT

BACK

Their tracks are pointy and split in two. Notice how the sides of the tracks curve out.
Front tracks are larger than the back tracks. Dewclaws sometimes show.

Scat is oval-shaped, like jellybeans.

BOBCAT

LEADING TOE

DENT — LEFT FRONT

RIGHT BACK

They have four toes on each foot. Their tracks are half as big as mountain lion tracks and don't show claw marks.

Scat looks like mountain lion scat, but smaller.

MOUNTAIN LION

LEADING TOE LEFT FRONT

DENT

RIGHT BACK

Their tracks are bigger than a coyote's. They have four toes on each foot. Their claws do not make marks in their tracks.

Mountain lions scrape dirt and twigs into a pile then urinate on it. This is called a scrape.

ABOUT the AUTHOR and ILLUSTRATOR

This is **GARY D. ROBSON'S** twentieth book. He lives in Montana near Yellowstone National Park, where he and his wife own an independent bookstore. He has written hundreds of articles and taught courses on various science and technology topics.

www.whopooped.com

ROBERT RATH is a book designer and illustrator living in Bozeman, Montana. Although he has worked with Scholastic Books, Lucasfilm, and Montana State University, his favorite project is keeping up with his family.

www.robertrath.net

OTHER BOOKS IN THE
WHO POOPED IN THE PARK?™
SERIES:

Acadia National Park

Big Bend National Park

Black Hills

Colorado Plateau

Glacier National Park

Grand Canyon National Park

Grand Teton National Park

Great Smoky Mountains National Park

North Woods

Olympic National Park

Red Rock Canyon National Conservation Area

Rocky Mountain National Park

Sequoia/Kings Canyon National Parks

Shenandoah National Park

The Sonoran Desert

Yellowstone National Park

Yosemite National Park